FARMING THROUGH THE AGES
IN PICTURES

The last of the steam ploughmen, at the Royal Show in 1954.

FARMING
THROUGH THE AGES
IN PICTURES

Robert Trow-Smith

FARMING PRESS

First published 1978
First paperback edition 1993

ISBN 0 85236 262 3

A catalogue record for this book is available from the British Library

Published by Farming Press Books
Wharfedale Road, Ipswich IP1 4LG, United Kingdom

Distributed in North America by Diamond Farm Enterprises,
Box 537, Alexandria Bay, NY 13607, USA

Cover design by Paul Young
Typeset by Galleon Photosetting, Ipswich
Printed and bound in Great Britain by Butler & Tanner Ltd, Frome and London

Contents

Introduction

To my great-grandfather, and to most of yours, the only trade he knew was farming; and the Britain that he knew was one expanse of farmland. True, there were many small, old towns; and some new towns were growing fast. But, apart from London and a few other cities, the old towns were mainly the business centres of the local farmers; and the new ones were populated by men lately come from the plough, or at most only a generation or two removed from it.

Today the countryside has shrunk, and every year shrinks ever faster. The men who earn their living from the soil are in a small minority, and each year become fewer. Britain—at least, lowland Britain—is now a chain of urban and suburban communities separated by farmland which, when glimpsed from a motorway, is a strange thing to 95 per cent of Britons.

To some of these it is even repellant and slightly derisory: full of machines and animals and crops and habits of life completely alien to the familiar world of pavements and street lights and elbowing neighbours. It is empty, and rather frightening.

To others, and these are a growing number, farming Britain (or such as is left of it) has the fascination of nostalgia, of an idyll with Nature, of the satisfactions of self-sufficiency.

To such sympathetically minded onlookers this book is addressed, as a simple guide in many pictures and few words to how farming Britain grew up. And to the hostile, in the hope that it may bring some understanding of the oddities which flash by them at seventy miles an hour. And, not least, to farmers themselves, in the hope that it may tell them a few things they didn't know about how their forefathers went about their businesses.

Friars, Rushden, Herts.
July 1978

Robert Trow-Smith

Introduction to the Second Edition

I am indebted to the many readers of the first edition of this book for their comments. They ranged from the lady in Coggeshall, Esex, who recognised the winnower in photo 82 as her uncle's foreman, to inmate KO4964 of H.M. Prison, Dartmoor, who identified and named eight of the (now defunct) farms around the hill grazings illustrated in photo 22.

Thurlby, Lincolnshire

1
Origins

BECAUSE BRITAIN was one of the later areas of the Old World to become civilised, in the accurate urban sense of that word, it has been assumed that its farming was equally retarded in its development. The agricultures of Egypt, Greece, Rome and the ancient empires of the East have a recorded history that stretches back some thousands of years: Britain's farming is documented for barely one thousand.

But the work of archaeologists is now beginning to suggest that in their arable fields and "estate management" at least, Britain's prehistoric cultivators were not as primitive as has hitherto been thought. The prehistorian is pushing back the frontiers of agricultural darkness, and long-held theories are now being upset right and left.

It now seems reasonably certain that some time around 4,000 BC the pattern of land use in Britain changed radically. The last of the glaciers of the Ice Age covered Britain until about 12,000 BC. Then a few hunters followed the ice northwards, living on what animals they could catch and on what edible vegetation they could find. But from about 4,000 BC onwards the evidence is showing that wild cereals were being ground for flour; and that the soil was being deliberately broken for the sowing of crops of wheat and barley, which were stored through the winter. Conscious "farming" had begun.

Sheep and cattle, too, began to be kept in domesticated herds, controlled by man's first friend, the dog. This livestock was periodically gathered in the causewayed camps which still break the smooth outlines of the chalk hills of the south.

By about 2,000 BC the open land, and some forest land which was burnt and felled, had become covered with the so-called "Celtic fields", of square or irregular shape, whose outlines can still be traced in aerial photographs. When the Romans arrived in AD 43 the farming of lowland Britain seems to have become so extensive and complex that some form of land management must have been exercised on an estate or tribal basis.

The Romans occupied and exploited Britain much as the British occupied and exploited India nearly 2,000 years later: a small force of soldiers and civilians controlled the country and Romanised the upper strata of native society. But much of the farming remained unchanged: it continued to be pursued by a thinly Romanised peasantry in its old fields. Only on the estates of the Roman villas, around such new cities as York and Lincoln, and later on land newly won from the drying Fens, did a more sophisticated land use appear. On the arable side it was based on larger fields, broken by a heavy iron-shod plough; on the livestock side, great flocks of sheep (some perhaps under imperial management) and new types of imported cattle appeared, to add hides and wool to the tributes which this, the remotest part of the Roman Empire, paid to Rome.

1/2 The early farmer, like the most modern, needed to gather and corral his livestock for calving and lambing, treatment, selection for slaughter or for sale. He also had to protect them against marauders, both two- and four-footed. His corrals had to be made with local material. In the lowlands he made his enclosures with earth banks and ditches, as at Thomas Hardy's Ring at Dorchester (**1**).

2

In the uplands he built stone walls around his enclosures. This stone circle (**2**) near Chagford, on Dartmoor, served to contain both the stock and the homes of the little farming community.

3

3 This spectacular series of lynchets perpetuates the fields which the early farmer ploughed out of the Devon moors. As the ploughing of each field proceeded, the soil tended to move downhill to the limit of the arable enclosure; and over the years a step was formed between an upper and a lower field. Such lynchets date from prehistoric times to the mediaeval period, and sometimes later. They are best preserved on marginal land such as this, which later farmers have not found it worth their while to cultivate. On the better lowlands, centuries of continuous cultivation have usually destroyed them.

4 The square outlines of "Celtic fields" are still visible, from the air, in the palimpsest of tracks and spinnies on the South Downs behind Shoreham. Such fields were in use from the early centuries of arable farming until Roman times.

4

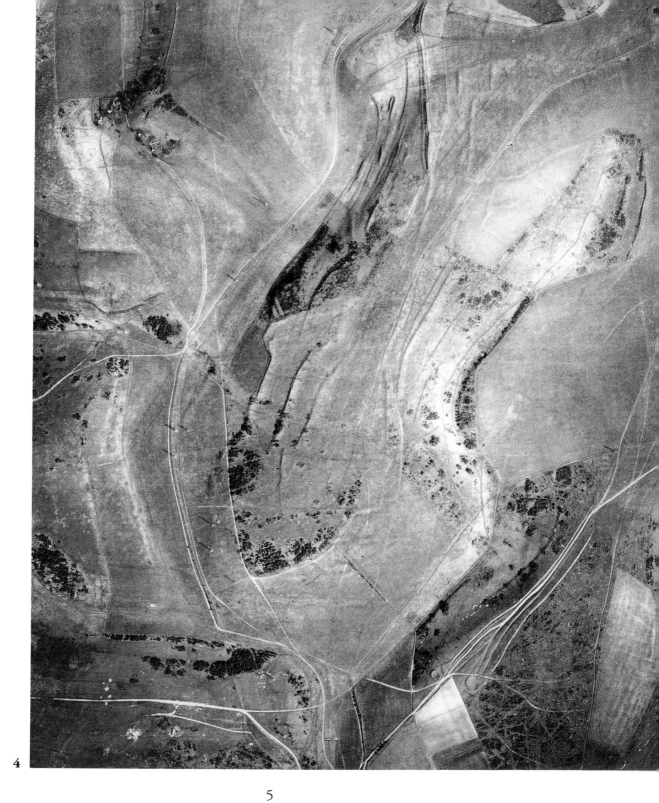

4

5 One of the most primitive implements of cultivation was the breast plough. Basically it was a pointed stick which was pushed through the soil to open a furrow for the seed. A similar tool was used to cut peat in the bogs of Scotland and Wales. The next stage in the evolution of the plough was a branch cut to the shape of a pull-hoe, which was drawn by the farmer through the ground. In time, cattle were harnessed to the branch, in place of man or woman. This was the prototype of the plough proper.

5

6 For 4,000 or 5,000 years, sheep have grazed around the prehistoric sanctuary of Avebury in Wiltshire. The downland of southern England was one of the first areas to be used by the Neolithic and Bronze Age farmer because it was free from woodland cover and grew short sweet grass for grazing. Its light soil was also easy to cultivate.

8

7/8 Only in the remotest parts of the British Isles do there remain the
descendants of the earliest domesticated stock of these islands. On St Kilda, the
most westerly of the Scottish islands and now uninhabited, the original breed
descended from the founding types of the Near East survived pure until the island
was evacuated early this century; and a few St Kildas (**7**) still remain there. Like
their distant cousins the Loaghten sheep of the Isle of Man (**8**) the rams carried
four horns, the ewes usually two. Today, these St Kildas and Loaghtens, and the
related Soays, are commonly found only in park flocks and in collections of rare
breeds of stock.

9

9 The prehistoric British farmer often stored his grain in pits in the ground, lined with basketwork to keep the corn clean. The more advanced Roman type of crop store was remarkably similar to the modern under-floor grain drier and store—a floor raised on stone staddles, below which air could circulate to keep the grain in good condition. This east granary at the Roman station of Corstopitum provisioned the garrison of Hadrian's Wall.

10 Bewick's engraving of a Chillingham "Wild or White Forest" bull shows an animal of 1789 from the herd still at Chillingham Castle, Northumberland. There is some reason to suppose that this white bovine stock was a Roman importation, perhaps to provision the garrison on the Wall.

10

2
Saxon and Mediaeval Times

WHEN THE Romans left Britain and the Saxon immigrants of the fifth and sixth centuries AD moved westwards and southwards, they found a countryside which was well populated and farmed—at least, in the kindlier lowlands. There is a growing body of evidence to disprove the old theory that they drove the Briton into Wales. It suggests that they did little to disturb the existing pattern of agriculture, and in fact occupied many of the villages and the farmsteads which were already there, avoiding only the country villas of the Romanised magnates. Indeed, they pursued the same pattern of farming, and merely, as the pressure of population grew, reclaimed more and more of the woodlands. Perhaps the only initial contribution of the Saxon settler to British agriculture was to make a heavier, more effective plough the common tool of the arable cultivator.

In time, the Anglo-Saxon–British peasantry and their masters began to evolve new ways of land use. Sub-division of holdings by inheritance through many generations produced a tangle of tiny arable strips; pasture for stock grew scarce as more and more land came under the plough; and, in the thickly populated Midlands and east, a common-field system of farming evolved. In this, the court of the manor dictated the cropping of the multitude of arable strips in the various fields of the village so that they could be cleared for the autumn pasturing of the village livestock. Only in the north and west, where the pressure on land was less, was this common field system rare or absent.

The system's great legacy to the English landscape was the ridge-and-furrow pattern of the fields of the Midlands, East Anglia and south-east England. Only now is modern deep ploughing obliterating the pattern of the strips of the manorial farmer of several centuries ago.

During the fourteenth century a succession of disasters destroyed a great deal of the peasant farming of mediaeval England. A deteriorating climate, with its attendant crop failures, and a series of epidemics of bubonic plague (of which the Black Death was merely the most severe) depopulated much of the countryside. The land which fell empty tended to be taken in hand and enclosed by the lords of the manors, to add to the considerable area of

12

11 Relics of the landscape of mediaeval England can be seen on the face of the modern countryside. At Kilby, in Leicestershire, the strips of the field system of the Middle Ages are revealed by the slanting sunshine in this aerial photograph taken just after World War II. Only the meadows alongside the stream bear no marks of ridge-and-furrow. The ploughing technique of the mediaeval farmer is illustrated in the S-lines followed by some of the strips: the length of the plough-team made it necessary to enter the plough into the bout and to leave it on the curve so that the team would have room to turn on the headland.

large individual fields they had already reclaimed from the manorial waste.

This chance to operate in larger units of land coincided with a boom in sheep; so that in the fifteenth century there was a perceptible change in the face of the farming landscape. The flocks of lords and prospering yeoman grazed a countryside which had once been a complex jigsaw of tiny arable strips. Much of the old common-field system remained, to die out almost completely in succeeding centuries; but a modern enclosed and hedged look had come upon the land.

Nonetheless, an unsophisticated subsistence farming survived, as islands of independence in the new rural empires. The Iron Age peasant of pre-Roman Britain would have found nothing wholly strange or incomprehensible in the materials and techniques of his Tudor counterparts.

12 Laxton in Nottinghamshire is one of a small handful of places to retain parts of its mediaeval open field system intact. An estate map of 1635 showed the few enclosed fields and the multitude of strips, often of half an acre, in the common fields. This shows a section of Mill Field at Laxton.

12

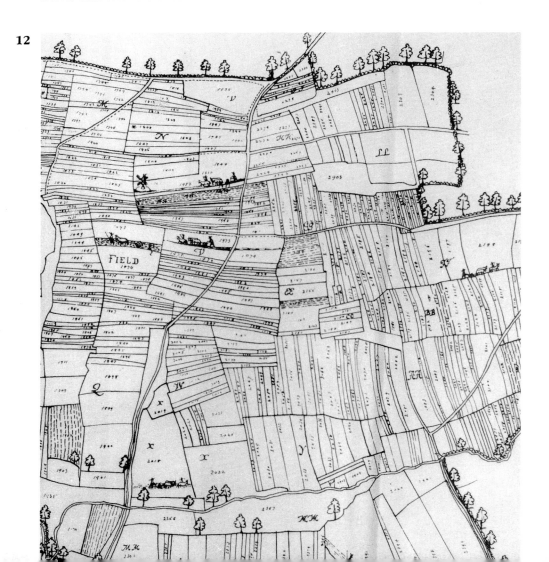

13 A mediaeval ploughing scene sketched by the scribe at the foot of a thirteenth century manuscript. The man in front uses a goad to encourage the team of four cattle which pull the plough (note the length of team and implement to be turned at the end of the bout; see **11**). The over-sized wheel is probably a figment of the artist's imagination; but the drawing accurately depicts the upright coulter behind the wheel which makes a vertical cut in the soil, the iron-tipped share which undercuts the furrow, and the mouldboard which turns over the furrow slice. The man behind is broadcasting seed into the furrows.

14 This Continental manuscript shows a different form of plough. The iron ploughshare at the end of the beam is tilted to move the broken soil sideways, instead of inverting the furrow slice as the mouldboard does. Note the modern type of collar on the draught beasts, which centred the strain on the shoulders.

14

15

15/16 A slightly developed form of the mediaeval plough was in use into the late nineteenth century. This implement, found in a Somerset barn, is believed to have been last used about 1850. It has the iron share at the base of the wooden mouldboard as in its ancestor (**13**). This design, which was certainly in use in Roman Britain, is reproduced in the modern iron plough, little altered.

16

17 One of the "works" which were performed by the servile tenants of the mediaeval manor as rent for their land held from the lord was the harvesting of the crop of the lord's own land. The manor reeve, or farm foreman, is supervising the tenant harvesters, who are cutting the corn high up the straw with sickles. A long length of uncut straw was probably left for grazing by stock which were pastured upon the stubble of the arable fields after the harvest had been carried.

18

18 The signs of old land use are everywhere to be seen by the discerning eye. As identification of deserted villages proceeds, lists of dozens in each of the midland and eastern counties grow yearly. They were abandoned as plague wiped out the villagers in the thirteenth century, or landlords enclosed the open fields for their own use, or pressure on land abated with recessions in agriculture. Lincolnshire is particularly rich in these deserted villages; and near Market Rasen a low hill which has never been cultivated bears the tell-tale signs of abandoned homesteads.

19 The homes of the mediaeval peasantry often lie only a few inches below the surface of the ground; and shallow excavation reveals the layout of farmhouses of six or seven centuries ago. This "longhouse" was discovered on Mr Michael Davidson's land at Harlington, Bedfordshire, in 1957. The two rooms of the house are plainly seen, one occupied by the family and the other by the livestock, with a connecting door. A few of these two-purpose homes are still in use in Wales and Western Scotland; and in Holland one roof still covers both family and stock in the new houses built on the reclaimed polders.

20

20 After the decline of the intensive subsistence farming of the mediaeval centuries, sheep became a major source of rural wealth. In the fourteenth century wool production and weaving were Europe's major industry; and in them the British flockmasters played a large part. Britain's monasteries in particular exploited the good natural grazing of their lands, and mediaeval export lists show that in Yorkshire alone the monastic houses exported the wool from nearly 200,000 sheep. One of the breeds held in greatest repute was the Cotswold, which fed on the short sweet grasses of their limestone hills and grew a fine fleece of high value. Their descendants, much altered in size and fleece type, still exist in a few flocks in Oxfordshire and Gloucestershire. These rams were in the very old flock of the Garnes of Aldsworth and Burford in 1820.

21 The hills of Wales were thinly sheeped by poor types of stock, eking out an existence among the rocks and bracken and rushes. Today the descendants of these mediaeval Welsh sheep are still to be found in remoter Wales. On Rhiw Mountain, in Caernarvonshire, this last flock of its type was photographed in 1952.

22

22 The old Welsh sheep in its purest form is probably to be found in the remote valleys of the hill land of Brecon. These at Abergwessin, at the end of the track from Tregaron to Beulah, had been brought down from the hills after lambing, for tailing and checking. In a bad season on these hills, a crop of only 40 lambs per 100 ewes is not uncommon, compared with over 200 per 100 ewes in breeds in kindlier counties. This lambing percentage has probably remained unchanged since mediaeval days.

23 A legacy of unpretentious architecture was bequeathed by the prospering yeomen who rebuilt the farmsteads of rural England around the turn of the sixteenth and seventeenth centuries. This typical small but exquisite example of the Great Rebuilding, as it has been called, is at Ashwell End, Hertfordshire—virtually unchanged for 400 years.

3
The Improving Centuries

IN THE four or five thousand years between the beginnings of farming in Britain and the end of the Middle Ages the techniques of agriculture changed slowly, usually imperceptibly, as one century followed another. But when the renaissance of European thought and art and commerce reached these islands, somewhat tardily, farming shared in the revolution.

New enquiring minds were brought to bear on the arts of the countryside (it was not yet a science), no less than on the arts of the stage, architecture and the written word. The result was an acceleration in the betterment of arable cultivation, of crops and—less marked at first—of livestock and its management. As always, the worst lagged far behind the best. The great improving landowners pioneered methods which the small subsistence farmer spurned for decades or centuries. But the best became better; and, more slowly, the worst became better too.

The hands in which improvement lay were largely those of the new yeomanry whose acreages were growing, whose tenures were more secure, whom education was now reaching in the new grammar schools, and whom prosperity was housing more elegantly. As the late sixteenth century Devon yeoman Robert Furse wrote in his diary: "Although our progenytors and forefathers were . . . but plene and sympell men and women and of smalle possessyon and habyte, yt have they . . . so run their corse that we are com to myche more possessyones credett and reputasyon than ever anye of them hadde."

Perhaps it was the new technical literature which played the greatest part in the application of knowledge to the making of money from crops and stock. As early as 1523 a Derbyshire squire, Fitzherbert, described in simple detail in his *Boke of Husbandry* the systems of stock and crop management which he had proved himself. Half a century later the East Anglian farmer Thomas Tusser published one of the classics of the countryside. His *Five Hundreth Points of Good Husbandry* enshrined in memorable doggerel the best of the agrarian practice of his day, for all to read. His pages are bespattered with phrases which have passed into common speech, from "Feb, fill the dike" to "sweet April showers that spring May flowers." Tusser

24

remained the chief mentor of the yeoman until Walter Blith published the book which, perhaps more than any other, set the farmer thinking. His *English Improver Improved* of 1652 contained advice—upon manuring, the use of grassland, drainage, new crops such as clover and lucerne, and the design of the plough—which could still be read with profit a century later.

As always, some of the new ideas were impracticable. They had to be refined by time and experience. Odd designs for dibbing in wheat, in place

¶ OCTOBER'S ABSTRACT.

CHAP. XVII.

1. Lay dry up and round,
   For barley, thy ground.
2. Too late doth kill,
   Too soon is as ill.
3. Maids, little and great,
   Pick clean seed wheat.
   Good ground doth crave,
   Choice seed to have.
   Flails (*b*) lustily thwack,
   Lest plough-seed lack.
4. Seed first, go fetch,
   For edish, or etch.
   Soil perfectly know,
   Ere edish ye sow.
5. White wheat, if ye please,
   Sow now upon pease.
   Sow first the best,
   And then the rest.

24

24 Thomas Tusser's *Five Hundreth Points of Good Husbandry*, first published in 1557 as *A Hundreth Points*, taught basic sound farming in memorable doggerel, much of which has passed into modern English phraseology. This page from a reprint of 1812, one of very many, gives a sample of his style of instruction. It was bred by mediaeval poetry out of the classical agronomists.

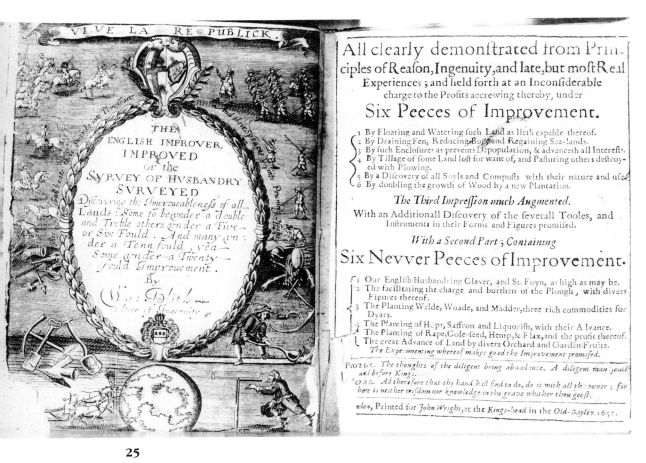

The title page reads:

VIVE LA RE-PUBLICK.

THE
ENGLISH IMPROVER
IMPROVED
or the
SVRVEY OF HVSBANDRY
SVRVEYED
Discovering the Improveableness of all
Lands :: Some to be under a double
and Treble others under a Five
or Six Fould . And many un-
der a Tenn fould , yea
Some under a Twenty
fould Improvement .
By
Wa: Blith
a lover of Ingenuitie

All clearly demonstrated from Prin-
ciples of Reason, Ingenuity, and late, but most Real
Experiences ; and held forth at an Inconsiderable
charge to the Profits accrewing thereby, under

Six Peeces of Improvement.

1. By Floating and Watering such Land as lieth capable thereof.
2. By Draining Fen, Reducing Boggs and Regaining Sea-lands.
3. By such Enclosures as prevents Depopulation, & advanceth all Interests.
4. By Tillage of some Land lost for want of, and Pasturing others destroy-ed with Plowing.
5. By a Discovery of all Soyls and Composts with their nature and use.
6. By doubling the growth of Wood by a new Plantation.

The Third Impression much Augmented.

With an Additionall Discovery of the severall Tooles, and
Instruments in their Forms and Figures promised.

With a Second Part ; Containing

Six Nevver Peeces of Improvement.

1. Our English Husbandring Claver, and St. Foyn, as high as may be.
2. The facilitating the charge and burthen of the Plough, with divers Figures thereof.
3. The Planting Welde, Woade, and Madder, three rich commodities for Dyars.
4. The Planting of Hopr, Saffron and Liquorish, with their Advance.

The Planting of Rape, Cole-feed, Hemp, & Flax, and the profit thereof.
The great Advance of Land by divers Orchard and Garden-Fruits.
The Experimenting whereof makes good the Improvement promised.

Prov. 21. 5. *The thoughts of the diligent bring abundance. A diligent man should*
stand before Kings.

Eccl. 9. 10. *All therefore that thy hand shall find to do, do it with all thy power ; for*
here is neither wisdom nor knowledge in the grave whither thou goest.

London, Printed for *John Wright*, at the Kings-head in the Old-Bayley. 1652.

25

25 The title page of Walter Blith's epic textbook, *The English Improver Improved* (1652). By assembling details of the best contemporary practices, and advocating new ones, Blith profoundly influenced the farming of several generations of English countrymen. The military scenes heading the sketches of rural work and tools reflected his bigotted Cromwellian faith.

of broadcasting the grain, led inevitably to Tull's invention of the seed drill. Men brooded upon the design of the plough, inspired by Blith's thoughts upon friction in the soil; and the days of an unwieldy mass of wood drawn by the brute force of a large ox-team were numbered. The potato, introduced in the late sixteenth century from America, was advanced as a crop which "might be propagated in great quantities as food for swine or other cattle."

The catalogue of new thoughts on old practices was endless, not least in

26 Jethro Tull (1674–1741), inventor of the seed drill. Tull was born at Basildon, Berkshire, and called to the Bar in 1724. He travelled Europe seeking ideas for his (unsuccessful) Berkshire farm, and took particular note of Continental practices in cereal growing. He was also an organist, and his two preoccupations led him to devise a machine with a hopper which fed grain through organ-like pipes with stops and sowed it in furrows in the ground, spaced at equal intervals (see **34**). The main idea behind this invention was to produce evenly spaced rows of plants between which a horse-drawn hoe could work to keep the crop clean—which was not possible with broadcast seed sown at random. His drill took nearly a century to win universal acceptance by his conservative neighbours.

26

livestock husbandry. Experiments began in earnest in the early eighteenth century into breeding better cattle, sheep and pigs; and these laid the foundations for the work of the famous stock improvers—Bakewell with his Longhorn cattle and Leicester sheep, the Quartley family with their Devon beef beasts, the Collings and their Scottish contemporaries with their dairy and beef Shorthorns.

To the urban eye the beginnings of revolution in the fields of Britain might still have passed unnoticed. What could not be missed was the great outward sign of the new rural affluence. Everywhere, bigger and more comfortable farmhouses were being built out of the local materials— Cotswold and Devon and Yorkshire stone, Midland and Kent oak and lime plaster, East Anglian brick. In the decades around 1600 the prospering yeomanry rehoused itself, enduringly and beautifully, leaving a legacy of unpretentious but superb architecture to future generations, unsurpassed anywhere else in the world.

27

27 Robert and Charles Collings (1749–1836), of Brampton, Yorkshire, were the designers of the Shorthorn breed of cattle which dominated the British livestock scene during the late nineteenth and early twentieth centuries. The original Shorthorn was a North Country beast evolved from crosses with imported Dutch cattle. From the Collings' new Shorthorn both the beef and the dairy Shorthorns were bred.

28 Lord Townshend (1674–1738), nicknamed Turnip Townshend, was one of the great landlords of Norfolk, with an absorbing interest in farming and politics. For long a forceful Whig statesman, he retired to his Raynham estate in 1730 after a quarrel with Walpole. He earned his nickname by his persistent advocacy of turnips as a field crop. Turnips had been grown in East Anglia since they had been brought in from the Netherlands in the sixteenth century, but Townshend's support made their merits better known as feed for stock in winter. When they were fed off in the field, the cattle and sheep both manured and consolidated the light Norfolk soils for later crops of cereals. Turnips also carried stock in better condition through the winter.

29 Thomas Coke, later Earl of Leicester (1752–1842), another great Norfolk landowner, began farming on his estate at Holkham in 1778. Like Townshend he was a passionate agriculturist and an even more skilful publicist of his ideas. He reclaimed and improved his coastal farms, exploited the Norfolk four-course rotation of wheat, roots, barley or oats, and a clover ley which came into nearly universal use, and bred improved types of Southdown sheep and Devon beef cattle. At his annual Holkham Shearings he entertained leading landowners and farmers from both home and abroad, and displayed to them his methods of crop and stock management.

29

30 Robert Bakewell (1725–1795), of Dishley Grange, near Loughborough in Leicestershire, was acclaimed as the originator of modern livestock breeding. He copied the techniques of contemporary bloodstock breeders, applied them principally to Longhorn cattle (**45**) and Leicester sheep (**46**) and by pseudo-secretive non-publicity caught the fancy of his time. His contribution to farming was not so much new types of stock he bred, which were failures in the long run in themselves, but more the impetus he gave to the adoption of techniques of line- and in-breeding and out-crossing.

30

31 John Ellman (1753–1832), of Glynde in Sussex, was a sheep improver and a founder of the Smithfield Show. He also helped Arthur Young, the great agricultural journalist, in his work. His improvement of Southdown sheep was the most important of his activities. He took the ancient short- and fine-woolled animal of his native South Downs and turned it from a long-legged and slow maturing sheep into a short-legged, meaty and early maturing mutton type. The greatest contribution of his new Southdown was as a crossing sire on other breeds, whose growth rate and early maturity were improved.

31

A
NEW INSTVCTION
OF PLOWING AND SET-
TING OF CORNE, HANDLED
IN MANNER OF A DIALOGVE
betweene a Ploughman and a
Scholler.

Wherein is proued plainely that Plowing and
Setting, is much more profitable and leſſe
chargeable, than Plowing and
Sowing.

By EDVVARD MAXEY. Gent.

He that withdraweth the Corne, the people will curſe him: but bleſſing
ſhall be vpon the head of him that ſelleth Corne. Prou.11.26.

Imprinted at London by *Felix Kyngſton,* dwelling in Pater
noſter Rowe, ouer againſt the ſigne of the
Checker. 1601.

32

32 One of the abortive ideas for sowing cereals which preceded Jethro Tull's drill, illustrated on the title page of Edward Maxey's *New Instruction* on ploughing and setting corn (1601). It used a perforated board which was moved across the field. The seed was dibbed through the holes, thus achieving the accurate and economical spacing of the Tull drill; but the method was unacceptably slow and tedious in operation. It probably never got further than the Tudor drawing board.

33 John Worlidge's seed drill, illustrated in his *Systema Agriculturae* of 1669, was an improved version of an earlier design by Gabriel Plattes. Like Tull's, his drill had a pipe with a regulated feed which led the seed from the hopper into a furrow made by a coulter. Further development of the design could have given this drill the honour which fell to Tull's drill 60 years later.

34 The design by Jethro Tull (**26**) of a seed drill was first published in his *The New Horse-Houghing Husbandry . . .* of 1731. The drill was an incidental, but essential, part of his technique of weed control in the cereal crop. It gave his horse-hoe the facilities to work down accurately spaced rows of grain; but Tull is remembered now for his drill rather than for his horse-hoeing husbandry.

34

Horſe-Hoing Husbandry :

OR, AN

ESSAY

On the PRINCIPLES of

TILLAGE and VEGETATION.

Wherein is ſhewn

A METHOD of introducing a Sort of *Vineyard-Culture* into the Corn-Fields,

In order to

Increaſe their Product, and diminiſh the common Expence;

By the Uſe of

INSTRUMENTS deſcribed in CUTS.

By *I. T.*

Cum Privilegio Regiæ Majeſtatis.

LONDON:

Printed for the AUTHOR, and Sold by *G. Strahan* in *Cornhill*; *T. Woodward* in *Fleet-Street*; *A. Miller* over-againſt St. *Clement's-Church* in the *Strand*; *J. Stagg* in *Weſtminſter-Hall*; and *J. Brindley* in *New-Bond-Street*.
MDCCXXXIII.

35 Jethro Tull's book gave fully detailed plans for the construction of the drill. This was, of course, before the days of patents.

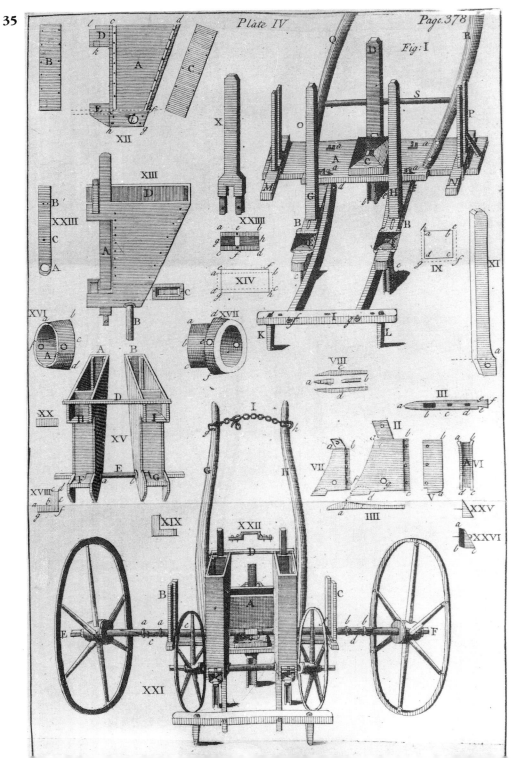

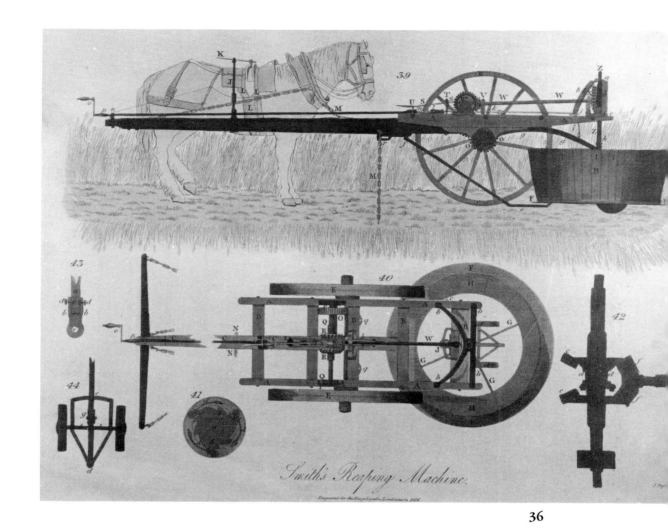

Smith's Reaping Machine.

Engraved for the Encyclopædia Londinensis 1816

36 The later eighteenth and early nineteenth century saw the growth of mechanical inventiveness, not least in agriculture. Smith of Deanston designed a rotary reaping machine with a horizontal circular cutter bar. This rotated at the base of a drum with sloping sides which moved the bottom of the straw against the cutting edge and then threw the crop into rows away from the feet of the propelling horse. Although it won several awards, Smith's reaper never went into commercial production.

37

37/38 In his *General View* of the farming of Essex (1807), Arthur Young illustrated this machine (**37**) which he had discovered in use on Norman Harding's farm at Hornchurch. It was designed to thresh the cereal crop without damaging the straw, which was in demand for thatching. The thresher was driven by a rotary horse-drive (**38**). This was declutched by foot to enable the sheaf to be withdrawn after the ears had passed between the threshing drums.

38

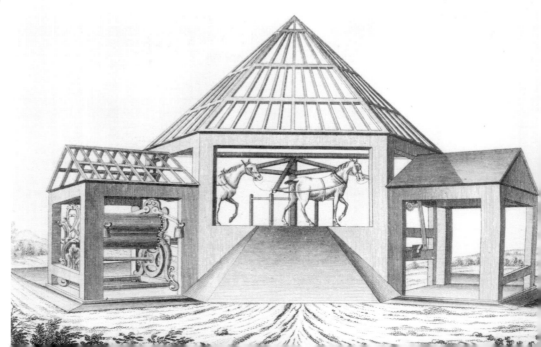

39 The rotary horse-drive shown in the preceding pictures was often housed in a circular building where the horses could pursue their day-long orbits, in all weathers, to drive the barn machinery before the invention of the internal-combustion engine. This example is at Lerryn, Cornwall.

39

40

40 The early livestock improvers of the eighteenth century had at hand a great array of local types of farm animals as raw material for their breeding experiments. Nearly every valley had its own type of sheep, nearly every village its parochial shape of pig. James Lambert illustrated in his *Countryman's Treasure* (1683) the stock of his time. The poverty of conformation reflects the poverty of artistry; but, none the less, much stock was poor and unthrifty.

41

41 Coke of Norfolk (see **29**) with one of his Holkham-bred Devon beasts. Derived from the foundation stock bred up by the Quartley family at Molland in north Devon, Coke's Devon was considered to be "the most perfect animal of its kind". Coke based his stock feeding upon his neighbour Townshend's turnip husbandry.

42 One of the most famous of the Collings brothers' new breed of Shorthorn was the bull *Patriot*, which was bought by a syndicate of Yorkshire breeders in 1810 for 500 guineas. It was bred by George Coates from his cow (right), which at 12 years old weighed 21 cwt and was 8 ft in girth. The shape of the new Shorthorn shows how breeders were concentrating the meat on the more expensive rear quarters, in place of the heavy-shouldered and small-rumped types of the past.

42

43 This bull was the famous *Comet*, sold for 1,000 guineas at the dispersal sale of the Collings' Shorthorn herd in 1810. It was the first thousand-guinea bull in farming history.

44 The great Shorthorn sire *Comet* calved in 1801 (see **43**) was a typical success story of eighteenth–nineteenth century livestock inbreeding. It was sired by "*Favourite*, dam by *Favourite* out of *Favourite's* dam." This is *Comet's* entry in the first volume of Coates Herd Book which lay, along with the Bible, on many hundreds of farmhouse tables.

45 A Longhorn cow of Robert Bakewell's type (see **31**). As with the Shorthorn, the meat had been moved to the rear and earlier maturity had been bred in; but in "improving" this old white-backed type of English cattle, Bakewell had also bred excessive fat into it and lost much of its milk. The breed, once nearly extinct, is now reviving.

46 One of Bakewell's new Leicester breed of sheep. As with his Longhorn cattle, Bakewell crossed upon local stock his own sires, often of other breeds, which made good the deficiencies in conformation of the female stock. He then fixed the improvement by close in-breeding. The new Leicester, although too fat and unprolific to succeed in itself, was used by the Culleys of Northumberland to found the Border Leicester breed, which revolutionised the production of prolific and milky ewes from hill sheep. This ewe was shown by the Duke of Bedford at Smithfield in 1799.

47 Improvement of the British pig was comparatively easy. The pig breeds twice a year, and the piglets reach breeding age well within the year—as compared to the annual breeding, long gestation and slow maturity of cattle. The domestic pig-keepers among Yorkshire weavers—who later turned their attention to breeding exotic flowers—were early and successful innovators of better strains. This "Yorkshire Hog" (1809) from Benjamin Rowley of Doncaster, weighed 12 cwt at four years old, was 9 ft 10 in long and 8 ft in girth. It toured the country to earn its owner nearly £3,000 in admission fees in three years.

48

48 A scene which epitomised the upper class view of rural Britain in the early nineteenth century. The squire—John Cotes, Member for Shropshire—inspects his farm staff at work, his dog at his horse's feet, attended by his bailiff, and his house in the background. The artist was Thomas Weaver (1744–1843), painter to the gentry: every scene guaranteed a bucolic idyll, every animal perfection.

49/50 The previous picture was an artist's idyll. These two drawings from late eighteenth and early nineteenth century books on farming were slightly closer to the reality of the time. They are from William Hogg's *Farmer's Wife* (1780) and Jeffreys Taylor's *The Farm* (1832).

49

50

4
Victorian and Edwardian Farming

IN VICTORIAN and Edwardian times the centuries of improvement turned into the centuries of accomplishment—accomplishment, that is, so far as the science, the engineering and (most vital) the economics of the period allowed.

During the seventy-seven years between the accession of the young Queen and the outbreak of World War I the fortunes of British farming had many ups and downs, not so much because of the internal inefficiency and the backwardness of the less progressive men but because of external circumstances. The first thing that happened to the industry—for industry it now was—was the repeal of the Corn Laws in 1846. This removed the protection from overseas competition which the arable farmer had enjoyed. By the 1880s the home market had become swamped by grain from the New World, where the plough, the seed drill and the railway followed the pioneers westwards across the great plains.

The British corn grower took several decades to climb out of the pit of insolvency into which this cast him; and it was not until the submarine blockade of 1917 that the nation considered him again and found it expedient to make his enterprise profitable. However, in spite of the overwhelming overseas competition, a slow improvement in the internal economics of cereal growing was made possible, partly by advances of science and engineering into the corn field; and partly by the removal of the smaller and less viable arable man into the production of milk for the vast number of new urban doorsteps.

For those who remained in arable farming, better drainage, new chemical fertilisers, improved strains of cereals, and labour- and time-saving machinery all helped to reduce production costs.

In 1843 John Reade invented a tile drainpipe which could be mass-produced cheaply enough to be laid by the million. In 1850 the Leeds engineer John Fowler made a mole plough which, pulled by one of his steam engines, drew a subterranean channel through the ground through which the surface water could escape. Gradually the fields of Britain became less sodden and more kindly.

Hitherto, the farmer had had to depend on natural manures for feeding his crops—the manure from his stock, dropped by animals grazing on grass or fodder crops, or carted from the stock yard; on green crops ploughed in; or on such industrial waste as shoddy from the textile mills. In the 1840s J.B. Lawes and J.H. Gilbert in Britain, along with the German chemist Liebig, began to experiment with, and soon to manufacture, chemical fertilisers for the balanced nutrition of the crop in the field. Soon these were substantially increasing the yields from the new strains of cereals, roots and grasses which the plant breeders were producing from their laboratories and trial plots.

These plant breeders were soon working not only on the traditional crops, but also on such new ones as the maritime beet from which the German scientist Marggraf had extracted sugar in 1747. Eventually this beet gave the British arable farmer a useful new root crop with which to rest and clean his land after a succession of cereal crops, as turnips did; but it was accepted so slowly here that while France and Germany each had 400 sugar-extracting beet factories by 1900, the first viable British factory was not established until 1912.

To handle these new and bigger crops grown with "bag" fertiliser on better-drained land, and to replace the labourers moving in their thousands into urban employment, the engineer applied his mind to the old problems of soil cultivation, harvesting and crop-handling, and to the motive power for them. The plough underwent a slow development as metal was substituted for wood in its construction, and the draught was eased by the reduction of the friction of the parts working in the soil. The horse, replacing the ox-team, remained the prime mover until it, in its turn, was superseded first by the steam engine and then by the internal-combustion-powered tractor.

The scythe and the sickle disappeared from the harvest field after Mr Cyrus McCormick of Chicago developed and patented a Northumberland invention and brought the mechanical reaper back to Britain for the Great Exhibition of 1851. The army of hand-reapers in the corn field disappeared almost overnight; and so did the brawny band of threshers who had, for untold centuries, beaten the grain out of the ear with their flails. They succumbed to the mechanical thresher which had slowly evolved through the eighteenth century and achieved efficiency and popularity in the first half of the nineteenth.

The farm tractor, today's most familiar country sight, was slower in acceptance than in development; and the horseman and his team were not finally ousted until the years between the two World Wars. American pioneers successfully worked the Dakota wheatlands in 1889 with a Rumely steam engine which they had turned into a tractor by the installation of a gasoline engine. British development then took over, and culminated in the Albone and Sanderson tractors which, with the Marshall, dominated world markets before their factories were diverted to munition-making after 1914. But, ironically, the first tractor to win real popularity in Britain was the Fordson, of which Henry Ford sent over 6,000 for the ploughing-up campaign after October 1917.

Both the new machinery and the new breeds of livestock were seen by a far wider farming public through the agricultural shows which proliferated following the first Smithfield Club exhibition of 1799 and the Royal Agricultural Society's first show in 1839. A popular farming press also took such journals as *Farmer and Stockbreeder* into literally every farmhouse after the mid-nineteenth century.

Here farmers could read of, as well as see at the shows, the binders and reapers and tractors and milking machines; the perfected British breeds of beef cattle, the Devon and the Hereford and the Angus, which also stocked the pastures of the New World; the higher-yielding dairy breeds of Shorthorns and Ayrshires, and the imported Friesians which soon patterned every pasture black and white; the heavier and more prolific Down breeds of sheep; and the fecund and fast-fattening new types of pig, the Large and Middle Whites and the Saddlebacks and the coloured breeds.

Thus equipped, Britain and its farmers were plunged into war in 1914.

51/58 When the World Agricultural Congress was held in Paris in 1856, pictures of the main cattle breeds of Europe were made for a commemorative book, *Les Races Bovines*. These were drawn with anatomical precision by several French and Dutch artists, and they were probably the first accurate illustrations of live stock to appear in print. Among the stock of British origin or interest were: Ayrshire cow (**51**); Hereford bull (**52**); Durham bull (**53**); West Highland bull (**54**); Guernsey cow (**55**); Kerry cow (**56**); Devon bull (**57**); "Dutch cow" (**58**). All would be acknowledged by the breeders of today as being at least acceptable specimens of their breeds—a tribute to little more than half a century's previous work by the improvers in fixing the modern type.

51

52

53

54

55

56

59

59 Professor Robert Boutflour—Bobby Boutflour of Cirencester—had a shrewder eye than most for the practical cow. His commentary on this sketch of a "Dutch cow" of 1847 was "I have never seen its equal". From its published daily yields he calculated that it may have been the first 2,000-gallon cow in history. Note how this working milch cow, ancestor of the British Friesian, differed from the show type in the Paris book (**58**).

60

60 Early attempts to milk the cow mechanically were always ingenious, often comical, rarely effective. This Danish Neilsen machine was entered for the Royal Agricultural Society's trials of 1891 by the Newark firm of Nicholson. It won a silver medal; but in long-term use its over-strong suction was found to damage the cow's teats, and it never came into commercial sale. Another contemporary idea was to apply suction from the farmyard pump to the unfortunate cow's udder.

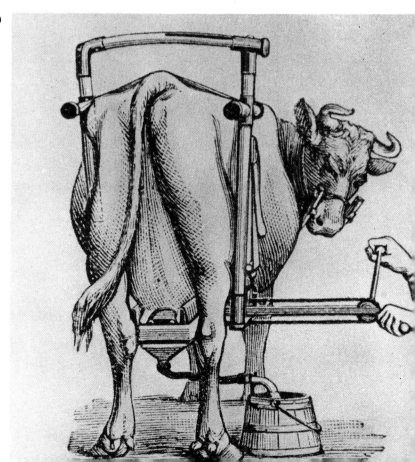

61

61 John Ellman of Glynde (**31**) evolved the modern Southdown sheep. Among the succeeding breeders who perfected it was Jonas Webb of Babraham, Cambridge. These three rams of his—Clumber, Liverpool and Woburn—won first prize at the Smithfield Show of 1842. The dark face of this early Southdown has been lost in the present British type, but is still retained in French Southdowns bred from early imports. Webb was pompous and overbearing to his inferiors (the artist, J.W. Giles, has caught the attitude) but obsequious to his betters. Typically, he named his rams after farming's upper crust.

62 The well-being of livestock was menaced in Victorian days by both disgruntled labourers and the "plague". The unrest among farm workers over their low wages, poor cottages and precarious jobs was aggravated by new labour-saving machinery which further threatened their employment. Not only did they break up the new-fangled threshers and reapers and fire the stacks; they also mutilated livestock. Societies sprang up everywhere to find and prosecute the culprits. The tail cutting of this notice was a mild example of their vengeance on their masters.

25 GUINEAS
REWARD.

Henfield Prosecuting Society.

WHEREAS some evil disposed Person or Persons, did, in the Night of Tuesday, the 8th Instant, break open the Stable on Furzefield Farm, in the Parish of Shermanbury, in the occupation of Mr. THOMAS PAGE, and maliciously CUT OFF and carry away

THE HAIR

FROM THE

TAILS OF 3 CART HORSES

the property of the said THOMAS PAGE.

A REWARD OF

FIVE GUINEAS

will be given to any Person or Persons giving Information of the Offender or Offenders, so that he or they may be Convicted thereof; such Reward to be paid by the Treasurer of the said Society, immediately after such Conviction.

THOMAS COPPARD, Clerk.

HORSHAM, 9th MAY, 1838.

A FURTHER REWARD OF

20 GUINEAS

will be paid on such Conviction as aforesaid, by me

THOMAS PAGE.

Printed by Charles Hunt, West Street, Horsham.

63 Growing imports of live cattle for slaughter, to feed the new millions of townspeople, brought European animal diseases into Britain as well. Foot-and-mouth disease arrived here in 1839, pleuro-pneumonia in 1841, sheep pox in 1847. After the arrival of the rinderpest from Eastern Europe in 1865—it was a virus-carried wasting fever—a Cattle Diseases Prevention Act restricted the import of stock and made the slaughter of suspect animals compulsory. Within a year the disease was stamped out, but only after 233,699 head of stock had been lost—including the seventy-six touchingly commemorated on this Cheshire gravestone.

63

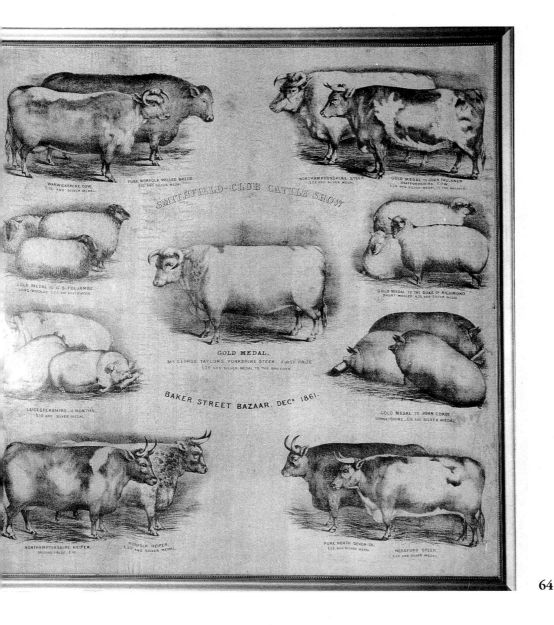

Within the illustration:

WARWICKSHIRE COW.
£10 AND SILVER MEDAL.

PURE NORFOLK POLLED BREED.
£10 AND SILVER MEDAL.

NORTHAMPTONSHIRE STEER.
£25 AND SILVER MEDAL

GOLD MEDAL TO JOHN FAULKNER
STAFFORDSHIRE COW.
£25 AND SILVER MEDAL TO THE BREEDER.

SMITHFIELD-CLUB CATTLE SHOW

GOLD MEDAL TO G. S. FOLJAMBE.
LONG-WOOLED £25 AND SILVER MEDAL

GOLD MEDAL TO THE DUKE OF RICHMOND
SHORT-WOOLED £25 AND SILVER MEDAL

GOLD MEDAL.
Mr GEORGE TAYLOR'S YORKSHIRE STEER. FIRST PRIZE
£25 AND SILVER MEDAL TO THE BREEDER

BAKER STREET BAZAAR. DECr 1861.

LEICESTERSHIRE. 11 MONTHS.
£10 AND SILVER MEDAL

GOLD MEDAL TO JOHN COATE
DORSETSHIRE £10 AND SILVER MEDAL

NORTHAMPTONSHIRE HEIFER.
SECOND PRIZE £10

NORFOLK HEIFER.
£25 AND SILVER MEDAL

PURE NORTH DEVON OX.
£25 AND SILVER MEDAL

HEREFORD STEER.
£25 AND SILVER MEDAL

64

64 This was a forerunner of the modern scarf illustrating Paris, or Venice, or the Costa Brava. In 1861, a trip by train to the Smithfield Show was the equivalent of today's farmer's fortnight in the Seychelles. The scarf shows the principal winners at the year's show at the Bazaar in Baker Street. It was the last of the shows at this horse mart. In 1862 the show moved to a new home at the Islington Agricultural Hall, where it stayed until World War II.

65

65 Sheep shearing in its traditional form at Nookton Farm, Durham, in 1896. The fleeces, after clipping by the team in the foreground, were rolled and sacked by the farm lads. Tar, heating in pans (right), was applied to cuts made by the shearers or to maggot wounds. Clippings were a neighbourly affair, from the fashionable Holkham Clipping of Lord Leicester to such parochial gatherings as this; and they ended in a large and liquid meal. These shearers, with their hand clippers, probably sheared eight sheep an hour each. With power-driven clippers the best modern shearer can handle up to 400 a day.

66 Richard Doyle drew for *Punch* for 16 years, and designed its famous cover, which he signed with his cipher of a "dicky-bird". His pencil poked fun at his contemporaries, and not least in his *Manners and Customs of ye Englyshe* of 1849. He was critical of the perennial cry that farming was ruined. This lampoon on a farmers' dinner, from *Manners and Customs*, was unsubtle, but justified at that time. Ruin had still to come.

MANNERS AND CVSTOMS OF Yᵉ ENGLYSHE IN 1849. Nᵒ. 38.

A BANQVET SHOWYNGE Yᵉ 'FARMERS FRIEND' IMPRESSYNGE ON Yᵉ AGRYCVLTVRAL INTEREST THAT IT IS RVINED.

67 John Bennett Lawes deserved his barrownetcy, however painful the pun referring to it in this cartoon. His trials on artificial fertilisers eventually led to the setting up of his superphosphate factory, the first bulk supplier of "bag manures" to the farmer; and to the establishment of the great research station at his home in Rothamsted, Herts. Initially his superphosphates were made by dissolving bones in sulphuric acid. In 1845 the discovery of coprolite beds nearby in Cambridgeshire provided his raw materials. Rumour had it that Lawes cared little where his bones came from; and it alleged that even battlefields were not sacred to his collectors.

67

68

68 One of the now lesser-known and under-honoured Victorian farming pioneers was William Somerville. In the 1880s an experimental farm at Cockle Park was set up by Northumberland County Council, and here Somerville sought cheap ways to improve the land which had fallen down to grass following the collapse of grain prices. He was successively professor of agriculture at Cambridge and at Oxford, and honorary director at Cockle Park. He evolved the system of heavy dressing with Lawes's supers and grazing the improving grasses hard, which improved both pasture condition and profits. He was the first of the pasture improvers who signposted the way for Stapledon at Aberystwyth.

69

69 The invention of the steam engine, and its application in a wheeled traction engine, was seen to have agricultural uses as early as 1810. In 1832 John Heathcoat, a Tiverton manufacturer and landowner, patented a system in which an "engine" drove cable drums which winched plough bodies across a field. It was not until the 1850s that John Fowler, the Leeds engineer, brought the system to the point of practicality: this engraving illustrates it. Steam ploughing swept the country. It had particular value on land too wet or heavy to carry a horse team or, later, a tractor. Steam ploughing was still in use after World War II: the last tackle belonged to George Patten, of Little Hadham, Herts.

70 Despite Fowler's steam tackle, traditional draught-team ploughing persisted. There was, indeed, no alternative on small fields until the tractor arrived. For countless centuries the "ox" was the universal draught animal. When the "ox" was a cow it could also be milked and breed calves; whether male or female, it could be fattened for beef at the end of its working life. On the small fields of Guernsey the island breed, for which an export sale for high quality milk had existed since the early nineteenth century, was still being yoked to the plough until World War I.

71 Ploughing with a single-handed Norfolk high-gallows plough in 1880. Such was the multiplicity of plough types, all favourites of a small locality or adapted to particular soil conditions, that Ransomes of Ipswich catalogued several hundreds of different shares and other fittings.

71

72/73 The age-old ways of corn harvest and haysel persisted in many places for half a century or more after mowers, binders and reapers came into common use. This picture (**72**) of a heavy and strong-strawed wheat crop being cut by hand was taken in 1900. The harvesters are using bagging hooks of the André type.

73

The above picture (**73**) shows the bagging hook in action, with a crook held in the left hand to gather the straw into the hook and then lower it in an unbound sheaf on to the stubble.

74/76 Stages in the evolution of the reaper. (**74**) In 1828, a quarter of a century before the American McCormick machine appeared at the Great Exhibition, the Rev. Patrick Bell of Forfar brought out a reaper with a moving canvas band to deliver the cut crop at the side in swathes. The idea of a pushed machine was abandoned when it was appreciated that pulling made better, and more comfortable, use of horsepower, and when side-delivery cleared a path for the horses so that the standing crop was not trodden. (**75**) With a side-mounted cutterbar, a side delivery device was not necessary to clear the swathe for the next bout. This Bamlett mowing machine of 1866, for both hay and corn, deposited the cut crop behind the cutterbar, but in order that the crop might be left in convenient sheaves for hand-binding by women and children a second man on the mower held a length of the standing crop on to the knife.

76

76 Wooden tines at the rear of the mowing machine in **75** combed the straw so that it lay all one way. In **76**, tined sails performed the same function mechanically.

77 "Mr McCormick's self-binder" which won a gold medal in the Royal Agricultural Society's trials in 1879. The tying of the cut sheaf on its way through the reaper from the cutterbar to the delivery canvas was made possible by the invention of the automatic knotting mechanism. This type of binder remained in use, basically unchanged, until the arrival of the first combine.

77

78 For women must rake and men must load! In this picture of 1905 women rake up the hay on a farm at Lockinge, Berkshire, for the men to pitch it on to the wagon with "pitch" forks. The wagons were being unloaded at the stacks being built behind. It was a technique of haymaking as old as wheeled vehicles themselves.

79

79 After the middle of the nineteenth century the threshing of the corn crop out of the stack was powered by the ubiquitous traction engine—here a Fowler "agricultural locomotive". It is driving a "double-blast" threshing drum and a jackstraw or elevator to restack the threshed straw. The two men on the rear stack are pitching the sheaves on to the platform of the thresher, where three more men cut the bands and feed the drum. The man at the rear of the thresher is bagging off the grain. Throughout the winter, from the 1870s to the 1930s, one of the traffic hazards in country lanes was the long convoy of this threshing tackle—traction engine, thresher and elevator, often steam-coal and water trucks, and the men's living caravan—moving from farm to farm.

80 In difficult hill country a low-slung cart made a high hay load more stable. Wagons similar to this Brecon haywain were still in use after World War II.

80

81 Before the thresher came into use, the corn crop was stacked into tall barns (rear), still common in lowland Britain, which had high double doors through which the wagons could pass. The unthreshed corn was stacked on one side of the central passageway in the barn, and thrown down into the passage between the doors for threshing, on canvas sheets, by men with flails. The threshed straw was then restacked on the other side, whence it was taken for bedding for the cattle in the open yard in front. Note the fine dovecote built into the gable-end of the granary. The farm is, appropriately, Dovecote Court, at Sollershope, Herefordshire.

83 Before the advent of the agricultural show, the great meeting place for the countryman was the large annual market. On a grand, indeed an international, scale these were held at Smithfield (this engraving is of the 1811 sales), where stock were brought on foot, often hundreds of miles, or by ship from the Low Countries. Absence of refrigeration on board made the importation of the live animal for slaughter on arrival a necessity; and the import of disease unavoidable (**63**). From the great Christmas sale at Smithfield in London evolved the Royal Smithfield Show, first held in 1799.

82 The threshed corn was then winnowed by throwing it up into the air from a basket similar to this, for the wind to blow away the chaff and leave the heavier grain to fall back on to a sheet. This winnowing basket, photographed at Grange Farm, Coggeshall, Essex, in 1940, was still in use for winnowing large flower seeds such as sweet peas.

84

84 The London Smithfield market was duplicated at scores of cattle and sheep fairs throughout Britain. This was the Boston May sheep fair, from a painting of about 1850. The sheep are all the local long-woolled Lincolns, now a rare breed.

85 The Great Exhibition of 1851 brought together probably the largest and certainly the most novel display of farm machinery and implements yet seen in Britain. It included not only Mr McCormick's new American reaper but also corn drills and threshers and elevators from Ransomes of Ipswich, barn machinery from Barretts of Reading, and Garrett's steam engines from Leiston—all world-renowned names in the history of farm machinery.

85

86 Since man can neither live nor work by bread alone, the means of making stimulating liquid refreshment was never far away—from the farm brewhouse with its ale for the harvest team to the travelling cider press, here at work outside the Cock Hotel at Bronllys, Brecon.

86

87

88

87/88 Two Victorian pioneers of farm mechanisation.

87 Bernhard Samuelson (1820–1905), the first president of the Agricultural Engineers' Association in 1875. Samuelson, half German, was a typical Victorian success story. He rose from apprentice to Member of Parliament, a baronetcy and the Privy Council. He entered upon the manufacture of farm machinery by buying James Gardner's Banbury turnip-cutter factory in 1848; and by 1872 he was selling 8,000 reapers and mowers a year, of his own design.

88 Robert Charles Ransome, once referred to by Gladstone as "that noble-minded gentleman". His grandfather founded the great Ipswich firm, later Ransomes Sims and Jefferies, in 1789. Here in 1856 Robert Ransome built the first successful steam plough for John Fowler of Leeds. He was the pioneer export salesman, travelling round the world to develop new markets for the products of his Ipswich factory. In a few years Ransome's implements were at work from Asia to South America.

5
The Inter-war Years

DURING THE Napoleonic Wars thousands of acres of marginal land which had not seen the plough since the rural depopulation of the fourteenth century came briefly into cultivation again. A hundred or more years later, during World War I, the same land was ploughed once more to feed a nation threatened with starvation, this time by the German submarine blockade. But yet again, when peace came and with it the return of imported food, this marginal land—and much which was far from marginal—fell down to rough grass.

Prices of bread and other grains had been guaranteed to the grower by the war-time Corn Production Act, and these were continued by the Agriculture Act of 1920. With the repeal of these Acts in 1921, when the war-time boom ended and reality returned, British arable farming declined steadily to the nadir of the 1930 depression. Few farmers stayed basically solvent: many survived solely on the credit from their bankers and merchants. Mechanisation, which had been a necessity when every spare body made munitions, became a luxury which few believed they could afford.

The agricultural sheet anchor of these inter-war years was the milk on the urban doorstep. This was a commodity so perishable that the British farmer was guaranteed the whole of his home market by the simple fact that the longest time that unrefrigerated milk would keep sweet was the few hours that elapsed during its train journey—from Somerset or Shropshire—from the cow to the consumer.

In the depression after the 1880s scores of thousands of men had turned from arable to dairy farming, led by the mass migration of ruined Scotsmen to grass farms within reach of London milk markets. Competition for the milk business with the big retail dairy companies became cut-throat: producers queued at distant office doors for contracts. Eventually, orderly marketing was achieved by a Milk Marketing Board, established under the Marketing Act of 1931. This had been sponsored by the increasingly powerful National Farmers' Union, itself an amalgam of the early 1920s of scores of local farmers' organisations. The same 1931 Act brought some order and fairer prices to other commodities—most notably pigs.

These and other measures helped the British farmer on his long haul back, first to solvency and, by the eve of World War II, to a reasonable prosperity. Science and engineering at last found a fruitful soil for growth on the large and medium-sized farm of the 1930s. By 1938 there were 60,000 tractors on British farms—a large proportion of them the little grey machines of that little grey, but explosive, genius, Harry Ferguson. The first combines were seen; and the days of the self-binding reaper, with its attendant army equipped with pitchforks and wagons, were numbered. By the end of World War II, corn growing had become an entirely mechanical process nearly everywhere.

Only the livestock man had perforce to pursue his old ways. Dairy farmers, it is true, were besieged by the adviser, the scientist and the engineer who showed them how to extract the milk from their cows in hygienic and mechanical ways and to cool the milk effectively. And while the pig was amenable to mass production, and eventually farm poultry, too, it was difficult—and indeed of dubious value—to mechanise the bullock and the ewe.

In any case, there was no short-term future for either, for both were slaughtered in large numbers when war again brought every possible acre under the plough to grow grain to feed a nation once more blockaded by the German submarines of World War II.

It is here that history may stop, and living memory take over the recent story of British farming. It had come an immense distance from the days when neolithic men sowed the first seed in the first man-made furrow and shepherded the first domesticated sheep.

89 In the small stone-walled fields of the Channel Islands there was an alternative to the Guernsey cattle as a work team (see **70**). These donkeys pulling a light single-furrow plough were also at work up to World War II.

90

90 On small farms there was a place for the mechanised plough. Indeed, its manoeuvrability made it the best inter-tree cultivator that the market gardener had ever had. The Douglas brothers bought this Wyles motorised plough in 1919; and up to World War II they calculated that they had turned over 15,000 acres with it on their land near Worthing, Sussex.

91 During World War II hundreds of thousands of acres which had seen a plough only twice since the Black Death—and that was in the emergencies of the Napoleonic and First World Wars—were cropped again. On each occasion, post-war shortages and prices encouraged the continuance of the work of reclamation and of winning new land for cropping. In particular, coastal marshes around the Wash and elsewhere were drained and broken up. This picture is of Mr Wallace Day's marshes at Fremington, Barnstaple, Devon, being ploughed, for the first time in history.

92/94 Horse teams, now seen only rarely in special classes at a few ploughing matches, succumbed slowly to the tractor during the inter-war years. These three pictures from the 1930s show them still at work: pulling a rib roll or clod crusher on the South Downs (**92**); an old horseman exercising his art of handling a team with a cultivator at Petworth, Sussex (**93**); and drilling spring wheat with a press drill at Coombes, Lancing, Sussex (**94**).

92

93

94

95

95 The sugar beet crop, although of high sale value, was expensive in hand work before it became fully mechanised. On average, 20 man-days an acre were spent on the crop, of which hoeing and singling the plants were the most demanding. Despite the invention of the precision drill and the mechanical singler, many acres are still singled by hand.

96 The first combines—or combined harvester, to use the original precise name—were seen at work in Britain in 1926. By 1941 there were one thousand in use. For the 1950 harvest there were 11,000 at work, including this early Massey-Harris harvesting wheat on the South Downs.

97 A scene in Rutland in 1920—and it could have been 1870, so long was the undisputed reign of the mobile steam outfit for threshing. This drum was a Clayton and Shuttleworth. It made use of a novel side-loading chute. The filled sacks of grain were being moved by sack barrow.

98 The baler, or baling press, first came on the market in 1882, under the name of Dederick's Perpetual Press. The following year the Royal Agricultural Society awarded a medal to a "straw trusser" from Howard's factory at Bedford. This used a binder-type knotting mechanism and took the straw straight from the thresher. For hay, most baling was done from the field stack or direct from the windrow in the field. This replaced the old system of using a hay knife to cut the hay from the stack, where it had been consolidated by weight, heat and moisture until it had the look and smell of cigar tobacco.

98

99 The thoroughbred horse needed hay made in thoroughbred ways. A high-priced trade demanded immaculate methods, from the time of cutting, through making in the field, to skilful stacking. "Rope and spar" classes at ploughing matches encouraged the making of straw ropes, drawn from the stack with a straw "twizzler", and of split or bent spars or pegs, both of which were used to secure the thatch on the stack. In this picture from Steane Park, Buckingham, Mr George Parrish was using twine in place of the old straw ropes to thatch a stack of thoroughbred hay.

100 Positively the "last appearances" of the heavy horse, once so ubiquitous on the British farm, are in classes at agricultural and horse shows. In the preparation of the entries the horseman exercises his last remaining art in the presentation of the mane, tail and feather. The scene is the judging at the Heavy Horse Show at Peterborough. Most of the remaining heavy horses are now owned by show breeders, breweries (for dray horses) and by American fanciers. The last heavy horses at work in significant numbers were on Fenland vegetable farms, where they were used for carting.

101

101 The dying livestock breeds of Britain have been saved from extinction in "rare breed" parks. The last of the Norfolk Horns were still to be found on farms up to World War II. These were at Howe Hall, near Saffron Walden, Essex. Formerly common on the light soils of East Anglia, the Norfolk Horns were derided by the improvers as "contemptible" and "a wretched sort"; but the gaunt and lean sheep was a prolific breeder, with lambing percentages of over 200 in the eighteenth century. This made it a fecund sire for the establishment of such crossing breeds as the Suffolk.

102 Another breed now remaining only in the hands of farm livestock collectors is the Gloucestershire. These cattle carried the distinctive white finch-back marking of very ancient origin. These Gloucestershires were in the Duke of Beaufort's park at Badminton just before World War II. One or two local farms also had a few. They gave milk of very high butterfat content, a trait probably fixed by breeders on West Country cheese farms.

102

103

103 The winter ration of bullocks housed in the traditional way, and even of dairy cows before "scientific" feeding, required four basic ingredients—hay, "cattle cake", chaff and such roots as mangolds and swedes. The chaff could be made out of the stack in the way illustrated here from Major J. A. Mansen's farm at Tring, Herts. The chaff cutter was evolved from a simple cutting box, which was a rectangular container with a guillotine knife at one end, worked by hand. By the end of the eighteenth century these were being made by most country craftsmen.

104 The traditional system of yarding beef cattle (see **81**) has now largely succumbed to the American-type beef lot with its mechanised feeding. The economics of yarding, as here at Britwell, Berkshire, between the wars, were always dubious. On paper, production costs exceeded returns; but livestock feeders persuaded themselves that sufficient profit was left in the farmyard manure which the beasts made, and that they used such residues of farm crops as straw. The cattle also ate the roots which had to be grown as part of the crop rotation. Therefore, they said, yarded beasts were an essential and not unprofitable part of their farming systems—and they may well have been right.

104

105 Roots for wintered stock were stored in a field clamp or in the barn, where they were sliced and mixed with chaff and broken cake. The rotary cage in the background cleaned the earth off the roots, which were then sliced by the root-cutter (left) and carted to the feeding troughs in the yard in the straw skips. The picture, taken in 1936, was from Manor Farm at Compton, Berks. Roots were criticised by the new breed of nutritional scientist as wasteful in handling because they contained 90 per cent water. But, as Professor Robert Boutflour once remarked, "What nutritious water!"

105

106 While the modernisation of milk production went on steadily in the lowlands, upland dairy farmers, and particularly those making the provincial cheeses, still milked by hand. Often, they did not even use a separator for the extraction of the cream but continued to skim the milk pans. This Wensleydale cheesemaker of 1934 had milked in the field—evidence the carrying strap on the churn—and made her cheese in earthenware moulds.

107

107 Alongside the upland dairy farm with its hand work, the innovating lowland producer was beginning to milk his cows mechanically at pasture, by means of one of Arthur Hosier's milking bails. This Wiltshire dairy pioneer had taken the new milking equipment and installed it in a mobile shed, with an engine-driven

vacuum pump to power the teat-cups. On far-flung grassland such as this at Liddington, on the Wiltshire downs, it was easier to take the machine to the cow than vice-versa.

108

108 Modern equipment on the sheep farm, until the arrival of custom-built races and mechanical clipping, went no further than a hayrack in the field. A Devon lamb waits expectantly.

109

109/110 On the arable sheep farm the flock had been hurdled since at least the sixteenth century days of Thomas Tusser. It was penned over a field of stubble or roots, the hurdles being moved a few yards every day. The keep was thus used with the minimum of waste by sheep which would otherwise range wide and choosily for their feed. The shepherd, perhaps with a boy, had the sole care of the flock; and at lambing time at least lived in a caravan alongside his ewes. The photograph below (**110**) is believed to be that of Shepherd Andrews, of East Hendred on the Berkshire Downs, in 1920.

111

111 Sheep dipping or washing in Wiltshire in the early 1920s. Their immersion in a dip of an arsenical solution controlled the scab, a highly infectious and debilitating disease of ancient origin. A handy stream could be used when it was intended merely to wash the fleece before shearing, to earn the higher price for clean wool.

112 Sheep have been housed in winter for many centuries, particularly on wet and heavy land. On the author's family's farm the Cistercian monks *rebuilt* the "schepehose" in 1485. After falling into centuries-long disuse in Britain, but not on the Continent, sheep housing was revived before World War II. This healthy, capacious and labour-saving house sheltered the Clun flock of Mr Henry Fell at Worlaby, Lincs.

112

113

113 During the two World Wars much of the work on British farms was done by members of the Women's Land Army, with pleasure to themselves and profit to the nation. The tea pot, inseparable from women even in the middle of a field, was produced for this happy WLA gang in East Sussex in June, 1941.

114 A grim reminder that the nation was at war again for the second time in little more than twenty years. A sentry mounts guard as Italian prisoners of war stack a load of barley—not that a guard was ever essential for Italian prisoners, happy to be out of the fighting and well fed and safe.

Acknowledgements

PICTURE REFERENCE NUMBERS

Museum of English Rural Life, Reading: 49, 50, 65, 66, 67, 68, 71, 72, 73, 78, 81, 85, 92, 93, 94, 95, 96, 97, 98, 103, 104, 105, 106, 107, 110, 111,113, 114.
British Museum: 13, 17.
Crown copyright: 4, 11.
Author's copyright: 6, 7, 8, 9, 18, 19, 21, 23, 79, 91, 99, 100, 101, 107, 108, 109, 112.
Author's collection: 1, 2, 3, 5, 10, 12, 13, 14, 15, 16, 20, 22, 24, 25, 26, 27, 28, 29, 30, 31, 32, 33, 34, 35, 36, 37, 38, 39, 40, 41, 42, 43, 44, 45, 46, 47, 48, 51, 52, 53, 54, 55, 56, 57, 58, 59, 60, 61, 62, 63, 64, 69, 70, 74, 75, 76, 77, 79, 80, 82, 83, 84, 86, 87, 88, 89, 90, 102.

Index

ROBERT TROW-SMITH was for twenty-five years a technical and political writer on the staff of the former *Farmer and Stockbreeder*, and was its last editor from 1967–71. He is the author of several books on the history of British and European agriculture, including the two-volume *History of British Livestock Husbandry*. For twenty years he and his wife ran a small sheep farm in north Hertfordshire.

Farming Press Books & Videos

Below is a sample of the wide range of agricultural and veterinary books published by Farming Press. For more information or for a free illustrated book list please contact:

Farming Press Books & Videos,
Wharfedale Road, Ipswich IPI 4LG, United Kingdom
Telephone (0473) 241122   Fax (0473) 240501

The Horse in Husbandry   ● JONATHAN BROWN

Photographs of horses working on farms from 1890 to 1950, with an account of how they were managed.

Ploughman's Progress   ● ALFRED HALL

An illustrated account of the history of ploughing and ploughmanship focusing in particular on the World Ploughing Organisation and its 'Olympics of the Plough'.

Tractors Since 1889   ● MICHAEL WILLIAMS

An overview of the main developments in farm tractors from their stationary steam engine origins to the potential for satellite navigation. Illustrated with colour and black-and-white photographs.

New Hedges for the Countryside   ● MURRAY MACLEAN

Gives full details of hedge establishment, cultivation and maintenance for wind protection, boundaries, livestock containment and landscape appearance.

Farming and the Countryside   ● MIKE SOPER & ERIC CARTER

Traces the middle ground where farming and conservation meet in cooperation rather than confrontation.

Pearls in the Landscape   ● CHRIS PROBERT

The creation, construction, restoration and maintenance of farm and garden ponds for wildlife and countryside amenity.

Henry Brewis

The six books by Northumbrian farming humorist Henry Brewis include three cartoon volumes, stories, poems, a diary and *Country Dance*, the story of a hill farm.

Farming Press Books is part of the Morgan-Grampian Farming Press Group which publishes a range of farming magazines: *Arable Farming, Dairy Farmer, Farming News, Pig Farming, What's New in Farming.* For a specimen copy of any of these please contact the address above.